COOKIE TIME

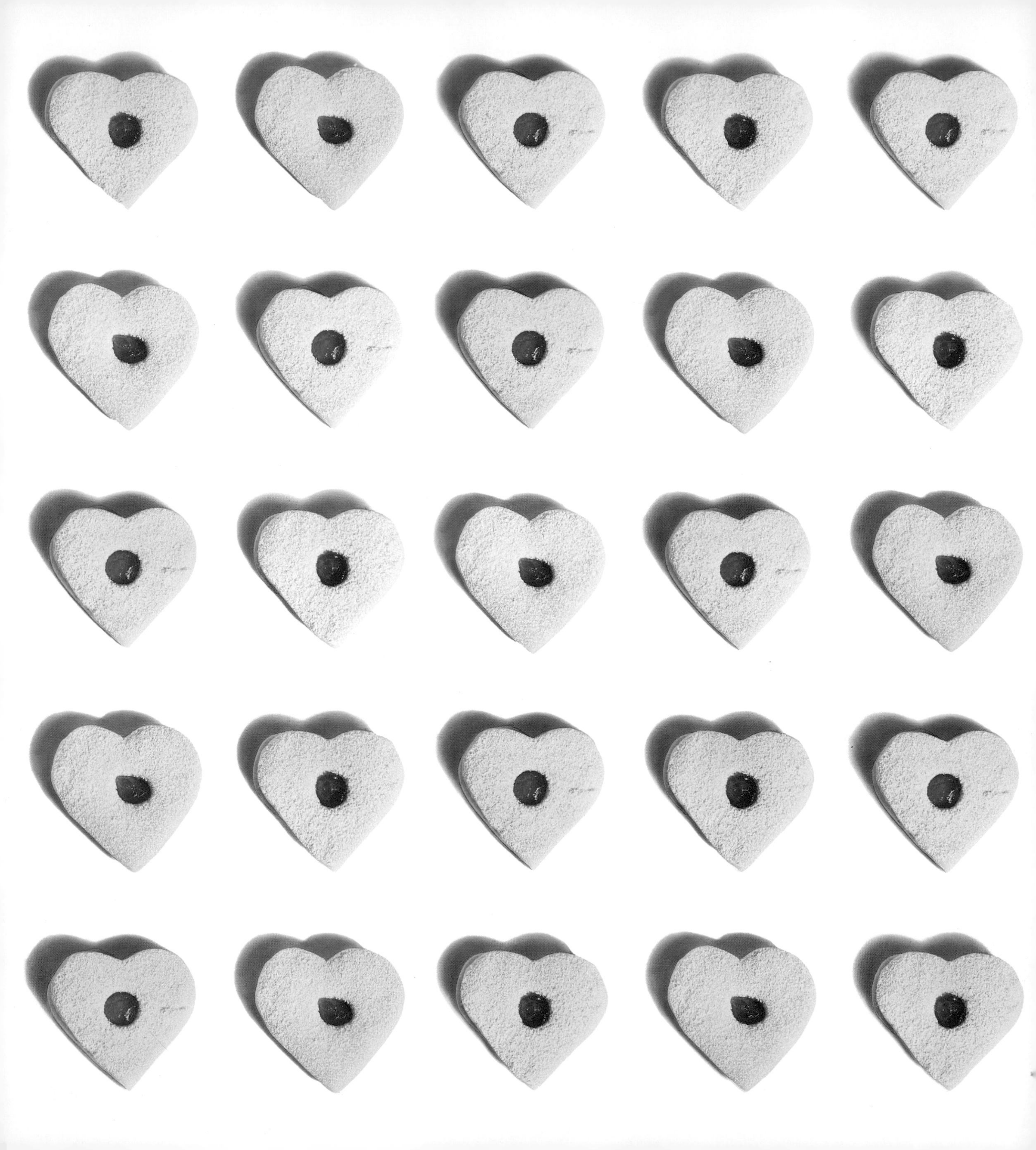

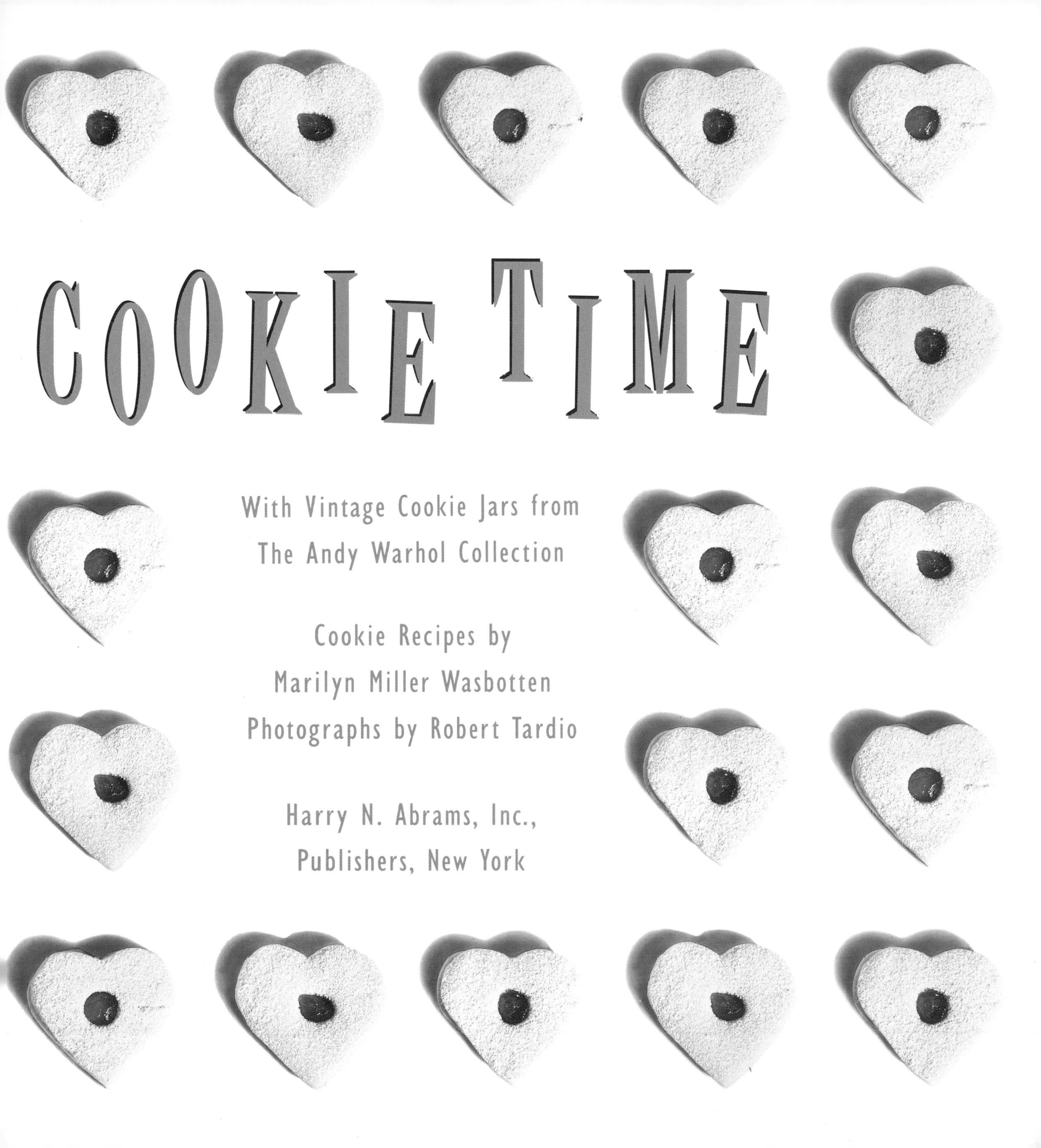

COOKIE TIME

With Vintage Cookie Jars from
The Andy Warhol Collection

Cookie Recipes by
Marilyn Miller Wasbotten
Photographs by Robert Tardio

Harry N. Abrams, Inc.,
Publishers, New York

PROJECT MANAGER: Darlene Geis
EDITOR: Ellen Rosefsky
DESIGNER: Joan Lockhart

The Andy Warhol Cookie Jars are now part of the private collection of Movado Watch Company, which made the collection available to be photographed. However, Movado Watch Company does not endorse and is not responsible in any way for the content of the book.

All recipes are suitable for kosher kitchens.

Library of Congress Cataloging-in-Publication Data

Wasbotten, Marilyn Miller.
Cookie time/recipes by Marilyn Miller Wasbotten;
photographs by Robert Tardio.
p. cm.
"Illustrated with cookie jars from the Andy Warhol Collection
owned by Movado Watch Company."
ISBN 0–8109–3173–7 (cloth)
1. Cookies. 2. Movado Watch Company. 3. Cookie jars—Private
collections—New York (N.Y.) I. Title.
TX772.W36 1992
641.8'654—dc20 91–22339
CIP

Published in 1992 by Harry N. Abrams, Incorporated, New York
A Times Mirror Company

Printed and bound in Japan

McCoy Chef's Head

CONT

ENTS

"...I asked Andy Warhol why he collects cookie jars. Andy whispered back, hiding a naughty-little-boy's smile, 'They're time pieces.' He bought many of his cookie jars in a Manhattan shop called Pieces of Time, which specializes in the simple, down-home household goods of the Thirties and Forties that are Andy's current passion."—Claude Picasso, 1972

INTROD

When the Warhol hoard was auctioned at Sotheby's in April 1988, not the least newsworthy items among the collectibles, jewelry, furniture, and paintings were the 139 vintage cookie jars, most of which the artist had picked up in the Manhattan flea market and sundry secondhand shops for a tiny fraction of the $247,830 they brought under John Marion's gavel. The lion's share of the pottery jars—35 of the 39 lots—was knocked down to Gerry Grinberg bidding on behalf of his Movado Watch Company.

Obviously there were many competing bidders eager to own an old cookie jar or two, especially if they belonged to Andy Warhol. But even before the auction, there were at least a thousand collectors of vintage cookie jars in the United States, a Cookie Jar Museum in Lemont, Illinois, and an *Illustrated Value Guide to Cookie Jars* by Ermagene Westfall (available from Schroeder Publishing Co., Paducah, Kentucky).

Early in this century, pottery cookie jars imitated the simple forms of tin biscuit canisters popular in colonial times. By the 1920s the potters had become more playful, and plump snowmen, cheerful Southern cooks, and lively World War I sailors ("send the boys some home-baked goodies") were offered by large pottery companies such as McCoy's.

As cookies, both home-baked and commercial, became a more popular American snack, cookie jar manufacturing came into its own. Favorite comic-strip characters like Popeye the

UCTION

Sailor Man, Disney's Mickey and Minnie Mouse, Dumbo, Donald Duck, and nursery characters like Goldilocks and Humpty-Dumpty were favorite subjects, lending themselves well to the rotund shapes and bright glazes of the pottery jars. The smiling kittens and pigs, Dutch boys and girls, Southern cooks and chefs are on the fine line separating pop art from kitsch. It is not surprising to find that they were appreciated by the leading pop artist of our time.

A selection of Andy Warhol's cookie jars was photographed especially for this book by Robert Tardio, with the kind permission of the Movado Watch Company. Since the jars were created with a purpose—to be filled with a stash of delicious cookies—a selection of some four dozen outstanding cookie recipes from Marilyn Miller Wasbotten, baker and owner of New York's Patisserie ChocoRem, is scattered throughout these pages.

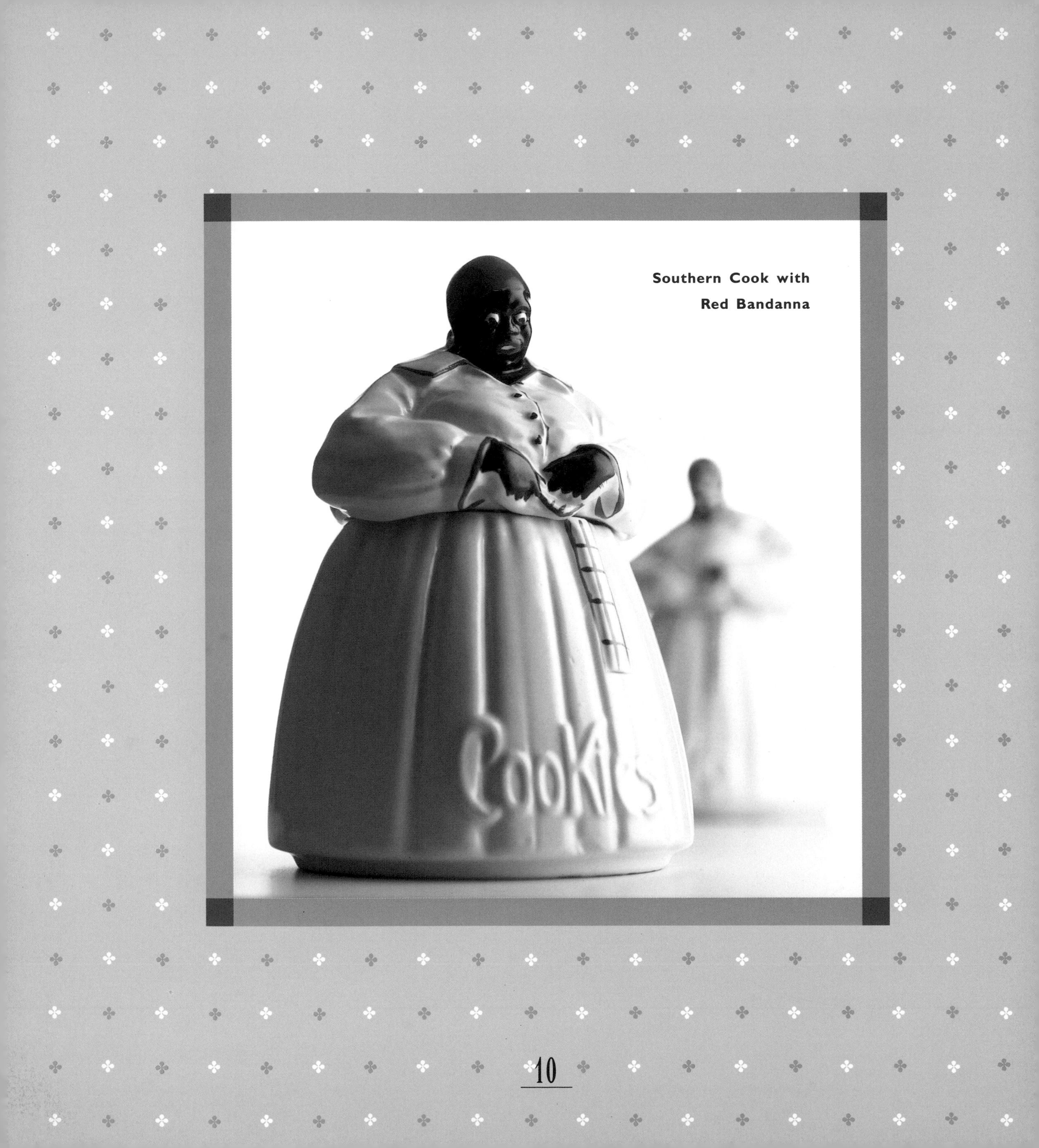

Southern Cook with
Red Bandanna

Aggie's Apricot Hearts

¾ cup butter
½ cup sugar
Zest of ½ lemon
Pinch of salt
2 egg yolks
2 cups flour
Apricot jam
Powdered sugar for sprinkling

Cream butter. Add sugar, lemon zest, and salt. Beat egg yolks until pale and frothy and add them. Add flour to mixture and mix well. Chill until firm. Roll out on floured board. Cut into heart shapes. Cut out hole in half of the hearts. Place on greased cookie sheet.

Bake at 350 degrees for 7 to 10 minutes. Spread whole hearts with apricot jam. Sprinkle other hearts with powdered sugar. Sandwich two cookies together so jam shows through hole.

Makes 36 cookies

Andy's Lemon Cheese Bars

CRUST

3/4 cup butter	**Pinch of salt**
1/2 cup sugar	**2 egg yolks**
Zest of 1/2 lemon	**2 cups flour**

Cream butter. Add sugar, lemon zest, and salt. Beat egg yolks until pale and frothy. Add yolks and flour and mix well. Chill until firm. Press into 9 × 13 × 2″ pan.

Bake at 350 degrees for 12 minutes.

FILLING

8 ounces cream cheese, softened	**Zest of 1 lemon, to taste**
1/3 cup sugar	**1/3 cup oil**
1 teaspoon lemon juice	**1 egg**

Beat cream cheese, sugar, lemon juice, lemon zest, oil, and egg until light and smooth. Spread over cooked pastry.

Bake at 350 degrees for 15 minutes. When cool, cut into bars.

Makes 15 bars

Dutch Boy

Anna's Chocolate Almond Drops

4 ounces almonds, slivered and blanched
½ cup butter or margarine, softened
5 tablespoons confectioners' sugar
⅔ cup flour, sifted
⅓ cup unsweetened cocoa

Grind ⅓ of the almonds until smooth and place in bowl. Repeat until all almonds are ground. Beat butter with confectioners' sugar until light and fluffy. Add remaining almonds, flour, and cocoa. Beat until dough leaves side of bowl. Cover and refrigerate until firm (about 2 hours). Drop by teaspoonfuls onto ungreased cookie sheet.

Bake at 325 degrees for 20 minutes. While still warm, roll in additional confectioners' sugar.

Makes 30 cookies

Rabbit in a Basket

American Bisque
Lady Pig

Aunt Bertha's Orange Raisin Cookies

2 1/4 cups flour
3/4 teaspoon salt
1 teaspoon baking powder
1 cup unsalted butter or margarine, softened
1 1/2 cups sugar
3 eggs
1 teaspoon vanilla extract
1 3/4 cups dark seedless raisins
3 tablespoons grated orange zest

Combine flour, salt, and baking powder in small bowl. With electric mixer, cream butter and sugar until well blended. Add eggs and vanilla; beat well (2 minutes medium speed). Add flour mixture. Beat just until flour is blended in. Add raisins and orange zest. Drop by tablespoonfuls onto greased cookie sheet.

Bake at 375 degrees for 10 to 12 minutes or until golden.

Makes 54 cookies

Red Brick Cottage

Aunt Jan's Pecan Wafers

¾ cup pecans, coarsely chopped
1 tablespoon butter
1 cup dark brown sugar
3 tablespoons flour
1 egg, beaten
1 teaspoon vanilla extract
¼ teaspoon salt

In a heavy pan, brown pecans lightly in butter. Add remaining ingredients, mixing well. Drop by teaspoonfuls onto well-greased cookie sheet dusted with flour. Leave 2 inches between cookies.

Bake at 350 degrees for 5 to 8 minutes. Remove from cookie sheet carefully while warm (two spatulas will help).

Makes 45 wafers

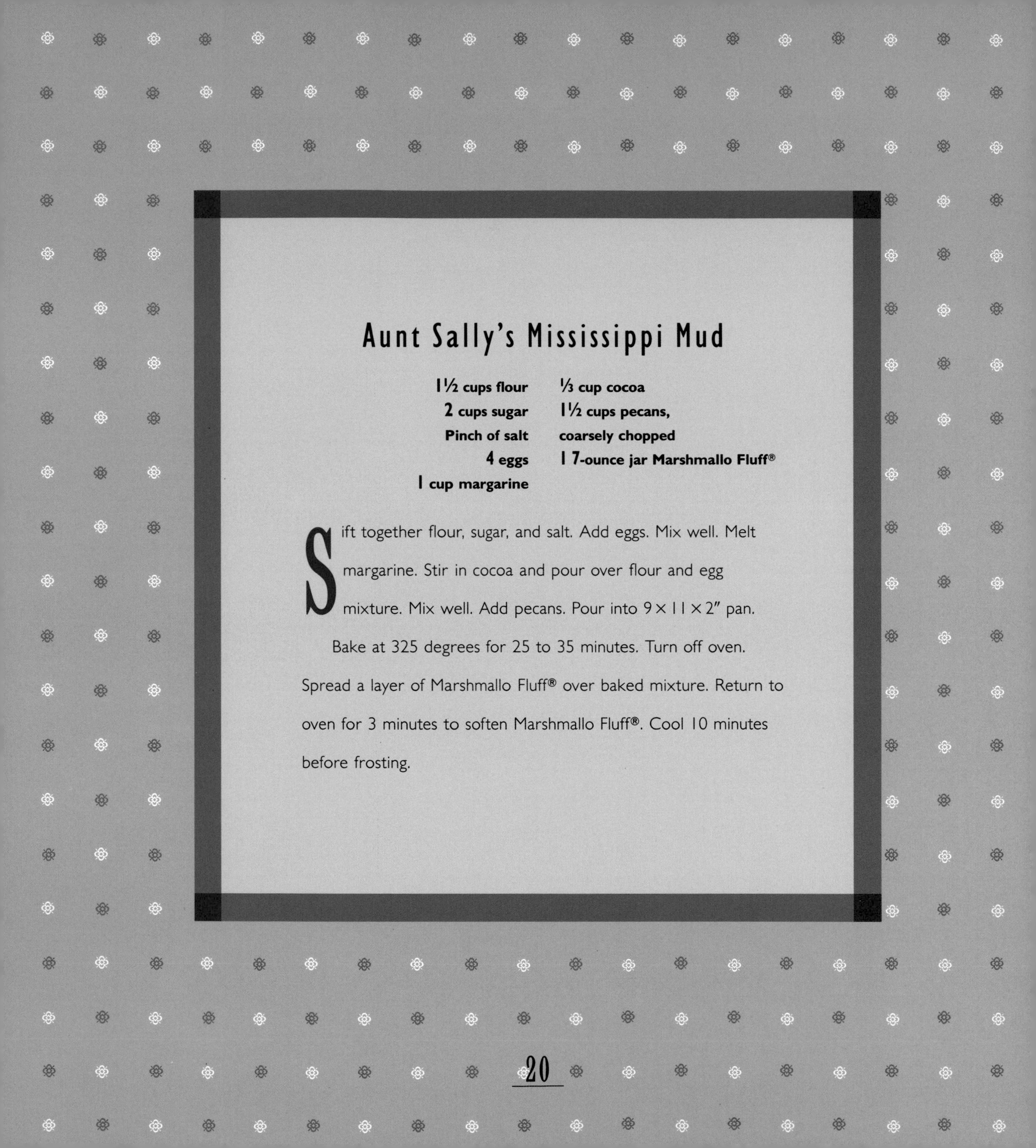

Aunt Sally's Mississippi Mud

1½ cups flour
2 cups sugar
Pinch of salt
4 eggs
1 cup margarine
⅓ cup cocoa
1½ cups pecans, coarsely chopped
1 7-ounce jar Marshmallo Fluff®

Sift together flour, sugar, and salt. Add eggs. Mix well. Melt margarine. Stir in cocoa and pour over flour and egg mixture. Mix well. Add pecans. Pour into 9 × 11 × 2″ pan.

Bake at 325 degrees for 25 to 35 minutes. Turn off oven. Spread a layer of Marshmallo Fluff® over baked mixture. Return to oven for 3 minutes to soften Marshmallo Fluff®. Cool 10 minutes before frosting.

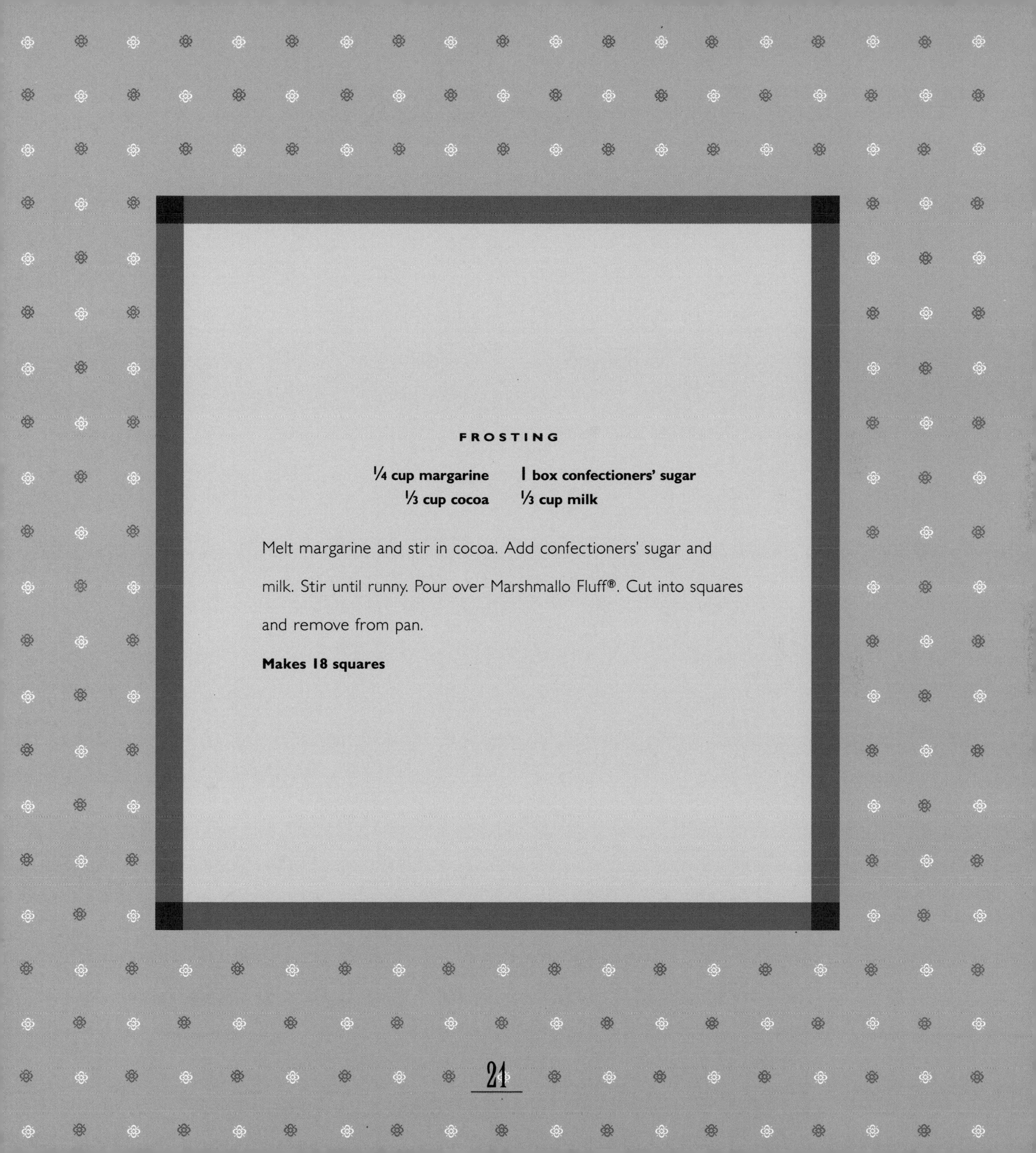

FROSTING

1/4 cup margarine	1 box confectioners' sugar
1/3 cup cocoa	1/3 cup milk

Melt margarine and stir in cocoa. Add confectioners' sugar and milk. Stir until runny. Pour over Marshmallo Fluff®. Cut into squares and remove from pan.

Makes 18 squares

McCoy
Clown Bust

Barbara's Orange Coconut Cookies

½ cup butter
¾ cup light brown sugar, firmly packed
1 egg
2 teaspoons grated orange zest
1 teaspoon vanilla extract
1¾ cups flour
2 teaspoons baking powder
½ teaspoon salt
½ cup sweetened coconut, shredded

Cream butter and sugar. Add egg, orange zest, and vanilla. Combine dry ingredients and add to creamed mixture. Mix well. Drop by teaspoonfuls onto ungreased cookie sheet. Bake at 375 degrees for 10 minutes.

Makes 72 cookies

Beryl's Boston Cookies

¾ cup flour
¼ teaspoon baking soda
Pinch of salt
½ teaspoon cinnamon
⅓ cup butter
½ cup sugar
1 egg, well beaten
⅓ cup walnuts, chopped
⅓ cup seeded raisins, chopped

Sift together flour, baking soda, salt, and cinnamon. Cream butter. Beat in sugar and egg. Stir in half of flour mixture. Add nuts, raisins, and remaining flour mixture. Drop by teaspoonfuls, 1 inch apart, onto greased cookie sheet.

Bake at 350 degrees for 12 minutes or until brown.

Makes 30 cookies

Two *Shawnee* Puss 'n Boots

Shawnee

Pirate

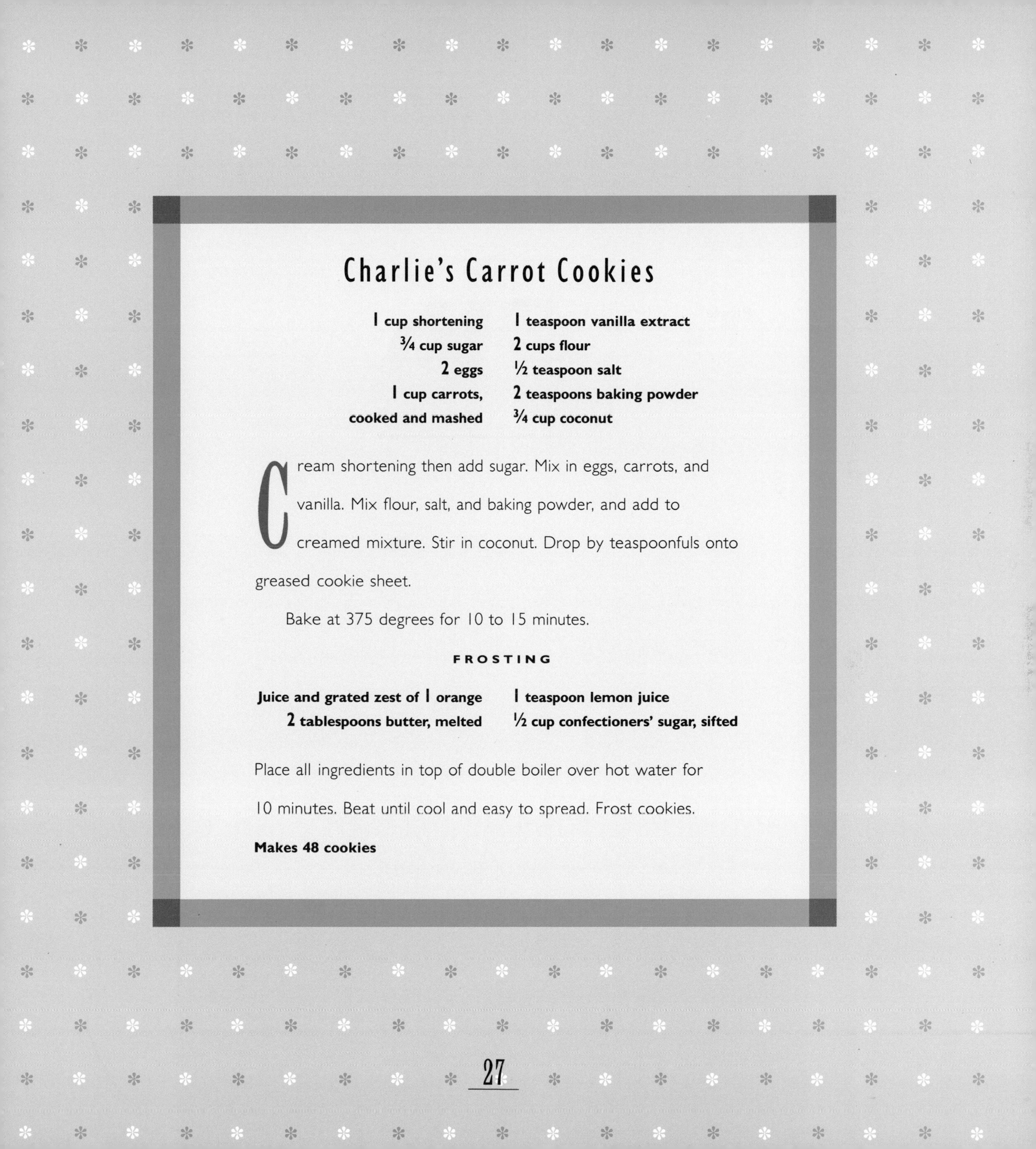

Charlie's Carrot Cookies

1 cup shortening
¾ cup sugar
2 eggs
1 cup carrots, cooked and mashed
1 teaspoon vanilla extract
2 cups flour
½ teaspoon salt
2 teaspoons baking powder
¾ cup coconut

Cream shortening then add sugar. Mix in eggs, carrots, and vanilla. Mix flour, salt, and baking powder, and add to creamed mixture. Stir in coconut. Drop by teaspoonfuls onto greased cookie sheet.

Bake at 375 degrees for 10 to 15 minutes.

FROSTING

Juice and grated zest of 1 orange
2 tablespoons butter, melted
1 teaspoon lemon juice
½ cup confectioners' sugar, sifted

Place all ingredients in top of double boiler over hot water for 10 minutes. Beat until cool and easy to spread. Frost cookies.

Makes 48 cookies

Man with
Mustache
and Goatee

Regal China
Boy with
Butter Churn

Daniel's Pecan Squares

CRUST

¾ cup butter	**2 egg yolks**
½ cup sugar	**2 cups flour**
Pinch of salt	

Cream butter. Add sugar and salt. Beat egg yolks until pale and frothy. Add flour and mix well. Press into 9 × 13 × 2″ pan.

Bake at 350 degrees for 15 minutes.

FILLING

2 cups butter	**2 cups light brown sugar**
½ cup honey	**8 cups pecan halves**
½ cup sugar	**½ cup cream**

Boil together butter, honey, sugar, and light brown sugar for 3 minutes. Remove from heat. Add pecan halves. Mix in cream. Spread on partially baked crust.

Bake again at 350 degrees for 30 minutes.

Makes 21 squares

Walt Disney Productions
Donald Duck

Regal China
Goldilocks

Darlene's Dainties

CRUST

¼ cup butter
4 ounces cream cheese
1 cup flour, sifted

Cream butter and cream cheese. Beat in flour slowly. Press the dough into lining of cups of mini-muffin tins to form cookie crusts.

FILLING

¾ cup brown sugar
1 teaspoon vanilla extract
1 egg
Dash of salt
¾ cup walnuts, chopped
1 teaspoon butter, melted
Confectioners' sugar for sprinkling

Blend together brown sugar, vanilla, egg, salt, walnuts, and butter. Spoon into crusts and bake at 350 degrees for 15 minutes. When cool, remove from muffin tins and sprinkle confectioners' sugar over cookies.

Makes 36 cookies

David's Chocolate Macaroons

2 large egg whites
¼ teaspoon salt
¾ cup sugar
½ teaspoon vanilla extract
1½ ounces unsweetened chocolate, melted
1½ cups sweetened coconut, shredded

Beat egg whites and salt until foamy. Add sugar, gradually. Beat until stiff peaks form. Add vanilla and melted chocolate. Fold in coconut. Drop by teaspoonfuls onto cookie sheet lined with parchment paper.

Bake at 350 degrees for 25 to 30 minutes.

Makes 36 macaroons

Dee's Molasses Delights

3/4 cup shortening
1 cup brown sugar
1 egg, beaten
1/4 cup molasses
1 teaspoon vanilla extract
2 1/4 cups flour, sifted
1/4 teaspoon salt
2 teaspoons baking soda
(dissolved in 1/2 tablespoon water)
1 teaspoon cinnamon
Pinch of ground cloves
1/4 teaspoon ginger
Granulated sugar
for rolling

Cream shortening and brown sugar. Stir in egg, molasses, and vanilla. Add the next six ingredients, stirring until blended. Chill for an hour. Shape into teaspoon-sized balls. Roll in granulated sugar. Press flat with fork dipped in flour. Place on ungreased cookie sheet.

Bake at 375 degrees for 6 to 8 minutes.

Makes 72 cookies

McCoy
"Hamms"
Bear

Diana's Spice Cookies

½ cup unsalted butter
½ cup confectioners' sugar
2 egg whites
1 cup flour, sifted
1 cup ground walnuts
¼ teaspoon cinnamon
¼ teaspoon ground cloves
¼ teaspoon nutmeg
¼ teaspoon allspice

Cream butter and sugar. Add egg whites. Mix flour, nuts, and spices. Add to butter mixture. Chill 2 hours. Roll on floured surface. Cut in desired shapes with cookie cutter.

Bake at 350 degrees for 10 to 12 minutes. When cookies have cooled, dip half of each in Tempered Chocolate.

TEMPERED CHOCOLATE

1 pound semisweet chocolate

Melt chocolate in double boiler. Put ¼ pound melted chocolate in refrigerator until stff. Pour remaining ¾ pound melted chocolate over cold chocolate. Stir until cool.

Makes 48 cookies

Donald's Chocolate Drop Cookies

½ cup shortening
1 cup dark brown sugar
1 egg
5 level tablespoons cocoa
½ cup buttermilk
1 teaspoon vanilla extract
1¼ cups flour, sifted
¼ teaspoon salt
1 cup walnuts, chopped

Cream shortening. Add brown sugar and egg. Mix thoroughly with next five ingredients, adding walnuts last. Drop by teaspoonfuls onto greased cookie sheet.

Bake at 350 degrees for 12 to 15 minutes.

Makes 36 cookies

American Bisque **Donkey with Cookies & Milk Wagon**

Cookie Jars on Parade

Ellen's Espresso Walnut Chocolate Chip Cookies

¼ cup unsalted butter, at room temperature
¼ cup shortening, at room temperature
½ cup dark brown sugar, packed
½ cup sugar
1 egg
½ teaspoon vanilla extract
1 cup flour
¼ cup cocoa
2 tablespoons instant espresso powder
½ teaspoon salt
½ teaspoon baking soda
½ cup walnuts, chopped
1 cup semisweet chocolate morsels

Cream together butter and shortening. Add brown and granulated sugar. Beat well and add egg and vanilla. Sift together other dry ingredients. Add to butter mixture and blend well. Stir in walnuts and chocolate. Drop by rounded teaspoonfuls 2½ inches apart onto cookie sheet lined with foil.

Bake at 350 degrees for 8 to 10 minutes until cookies are set and tiny cracks appear on top.

Makes 54 cookies

Elvira's Oatmeal Chocolate Chunks

1½ cups flour
½ teaspoon baking soda
1 teaspoon cinnamon
½ teaspoon salt
1 egg, well beaten
1 cup brown sugar
½ cup butter, melted
½ cup shortening, melted
¼ cup milk
1¾ cups quick cooking oats
½ cup raisins
½ cup walnuts, chopped
½ cup semisweet chocolate, chopped

Mix flour, baking soda, cinnamon, and salt. Add beaten egg and remaining ingredients. Mix well. Drop by teaspoonfuls onto ungreased cookie sheet.

Bake at 350 degrees for 10 to 12 minutes.

Makes 48 cookies

Tuggle Tuggle Tugboat with *Shawnee* Sailor

Grandma Janet's Cream Wafers

1 cup butter
⅓ cup whipping cream
2 cups flour
Granulated sugar

Mix first three ingredients well. Chill 1 hour. Roll out dough on lightly sugared surface. Cut in 2-inch circles. Sugar each side. Place on ungreased cookie sheet and prick each circle twice with fork.

Bake at 375 degrees for 7 to 9 minutes.

FILLING

¼ cup butter, softened
1 cup confectioners' sugar
1 egg yolk
1 teaspoon vanilla extract

Mix all ingredients well. Add a few drops of food coloring if desired. Sandwich two cookies together with filling.

Makes 30 wafers

McCoy
White Wood
Stove

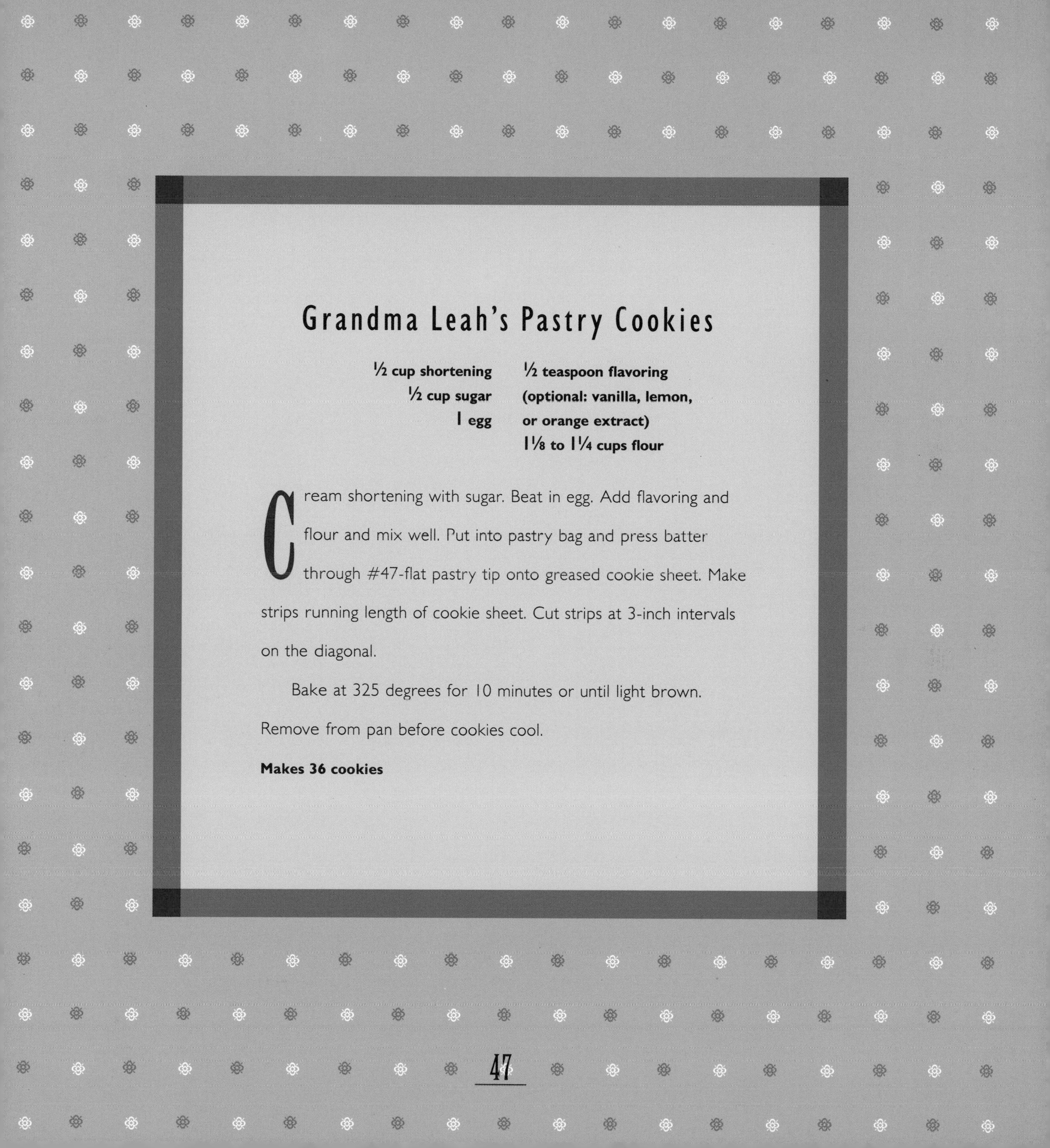

Grandma Leah's Pastry Cookies

½ cup shortening
½ cup sugar
1 egg
½ teaspoon flavoring (optional: vanilla, lemon, or orange extract)
1⅛ to 1¼ cups flour

Cream shortening with sugar. Beat in egg. Add flavoring and flour and mix well. Put into pastry bag and press batter through #47-flat pastry tip onto greased cookie sheet. Make strips running length of cookie sheet. Cut strips at 3-inch intervals on the diagonal.

Bake at 325 degrees for 10 minutes or until light brown. Remove from pan before cookies cool.

Makes 36 cookies

Joel's Chewies

2 eggs
2 cups dark brown sugar
1⅓ cups flour
Pinch of baking soda
1 cup nuts, chopped (walnuts or pecans)
1 teaspoon vanilla extract
Confectioners' sugar for dusting

Beat eggs and combine with brown sugar. Sift together flour and baking soda. Add to egg-and-sugar mixture. Add chopped nuts and vanilla. Pour into greased 9 × 13 × 2″ pan.

Bake at 350 degrees for 30 minutes. When cool, cut into squares and dust with confectioners' sugar.

Makes 15 cookies

Shawnee
Dutch Boy with Gold Buttons

Joni's Butter Pecan Melts

1 cup butter
⅓ cup confectioners' sugar
½ teaspoon vanilla extract
2 cups flour
1½ cups pecans, finely chopped
Confectioners' sugar for dusting

Cream butter. Add sugar and vanilla. Stir in flour until well blended. Add pecans. Shape into marble-sized balls.

Bake on ungreased cookie sheet at 300 degrees for 45 minutes. When cool, dust with confectioners' sugar.

Makes 48 cookies

Kara's Sesame Wafers

½ cup sesame seeds
1 tablespoon butter
1 cup dark brown sugar
3 tablespoons flour
1 egg, beaten
1 teaspoon vanilla extract
¼ teaspoon salt

In a heavy pan, brown sesame seeds lightly in butter. Add to the rest of the ingredients. Mix well and arrange by teaspoonfuls on a well-greased cookie sheet dusted with flour. Leave 2 inches between cookies.

Bake at 350 degrees for 5 to 8 minutes. Remove from pan carefully while warm (two spatulas will help).

Makes 45 wafers

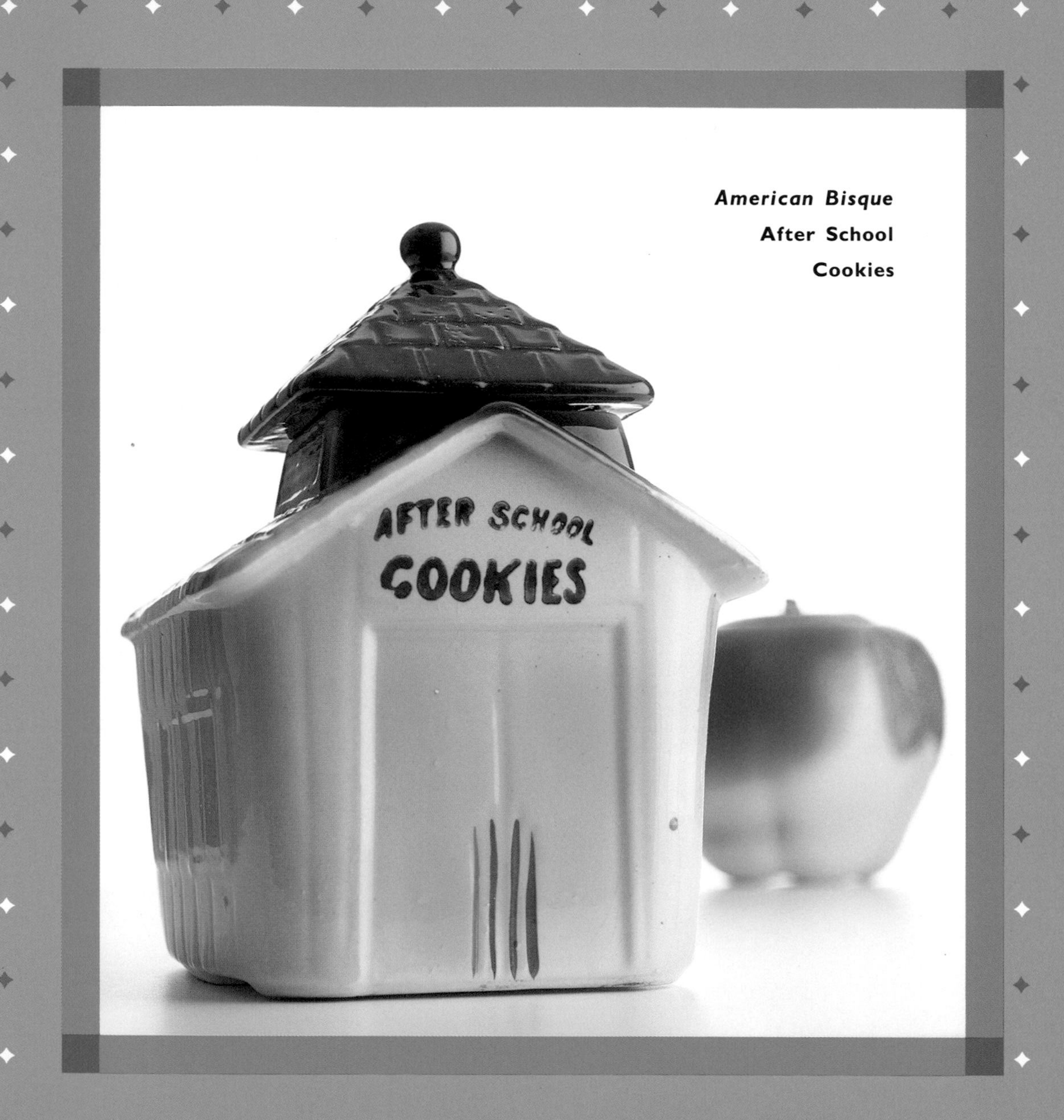

American Bisque
After School
Cookies

Kelly's Chocolate Pecan Kisses

¼ cup butter or margarine, softened
¾ cup sugar
1 egg
1½ squares (1½ ounces) unsweetened chocolate, melted
1½ teaspoons vanilla extract
½ cup flour
¼ teaspoon baking powder
½ teaspoon salt
1½ cups pecans, chopped

In a mixer, beat butter, sugar, egg, chocolate, and vanilla on medium speed. Mix well. Add flour, baking powder, and salt. Scrape down bowl with rubber spatula. Stir in pecans. Drop by rounded teaspoonfuls ½ inch apart on greased cookie sheet.

Bake at 350 degrees for 10 minutes.

Makes 48 cookies

King's Crescents

1 cup butter
½ cup plus 2 tablespoons confectioners' sugar
2 ¾ cups flour
½ cup walnuts, coarsely chopped
Powdered sugar for rolling

Cream butter. Add confectioners' sugar. Beat until light and fluffy. Add flour and walnuts. Shape into crescents.

Bake on greased cookie sheet at 350 degrees for 15 to 20 minutes or until golden around the edges. Roll in powdered sugar while warm.

Makes 55 cookies

Two *Cardinal* Soldiers

Kristen's Coconut Chews

¾ cup half shortening, half sweet butter
¾ cup confectioners' sugar
1½ cups flour
2 eggs
1 cup dark brown sugar, packed
2 tablespoons flour
½ teaspoon baking powder
½ teaspoon salt
½ teaspoon vanilla extract
½ cup walnuts, chopped
½ cup sweetened coconut

Cream shortening, butter, and confectioners' sugar thoroughly. Stir flour into mixture. Press into 9 × 13 × 2″ pan.

Bake at 350 degrees for 15 minutes. Combine remaining ingredients. Mix well. Spread over baked crust. Bake 20 minutes more. While warm, spread with icing.

ICING

1½ cups confectioners' sugar
3 tablespoons butter, melted
3 tablespoons orange juice
1 teaspoon lemon juice

Mix all ingredients together and spread on warm coconut chews.

Makes 30 cookies

Lisa's Linzer Cookies

1 cup butter
3/4 cup sugar
1 egg
1/4 teaspoon vanilla extract
1/2 cup filbert nuts, ground
2 1/2 cups flour
1/2 teaspoon cinnamon
1/2 teaspoon baking powder
Confectioners' sugar for sprinkling
Raspberry jam

Cream butter and sugar. Add egg and vanilla. Combine filberts, flour, cinnamon, and baking powder. Add to creamed mixture and mix well. Wrap in plastic and refrigerate at least 1 hour. For top cookie, roll out half the dough on lightly floured surface to 1/8-inch thickness. Cut out cookies with 2-inch cookie cutter. Cut small hole out of centers. For bottom cookie, roll out remaining dough. Cut out solid cookies with 2-inch cookie cutter. Place cookies 1/2 inch apart on ungreased cookie sheet.

Bake at 350 degrees for 12 minutes. Sprinkle confectioners' sugar over cutout cookies. Put a dollop of jam in center of solid cookies. Sandwich two cookies together so jam shows through hole.

Makes 72 cookies

Shawnee Dutch Girl with Blue Skirt and Dutch Boy

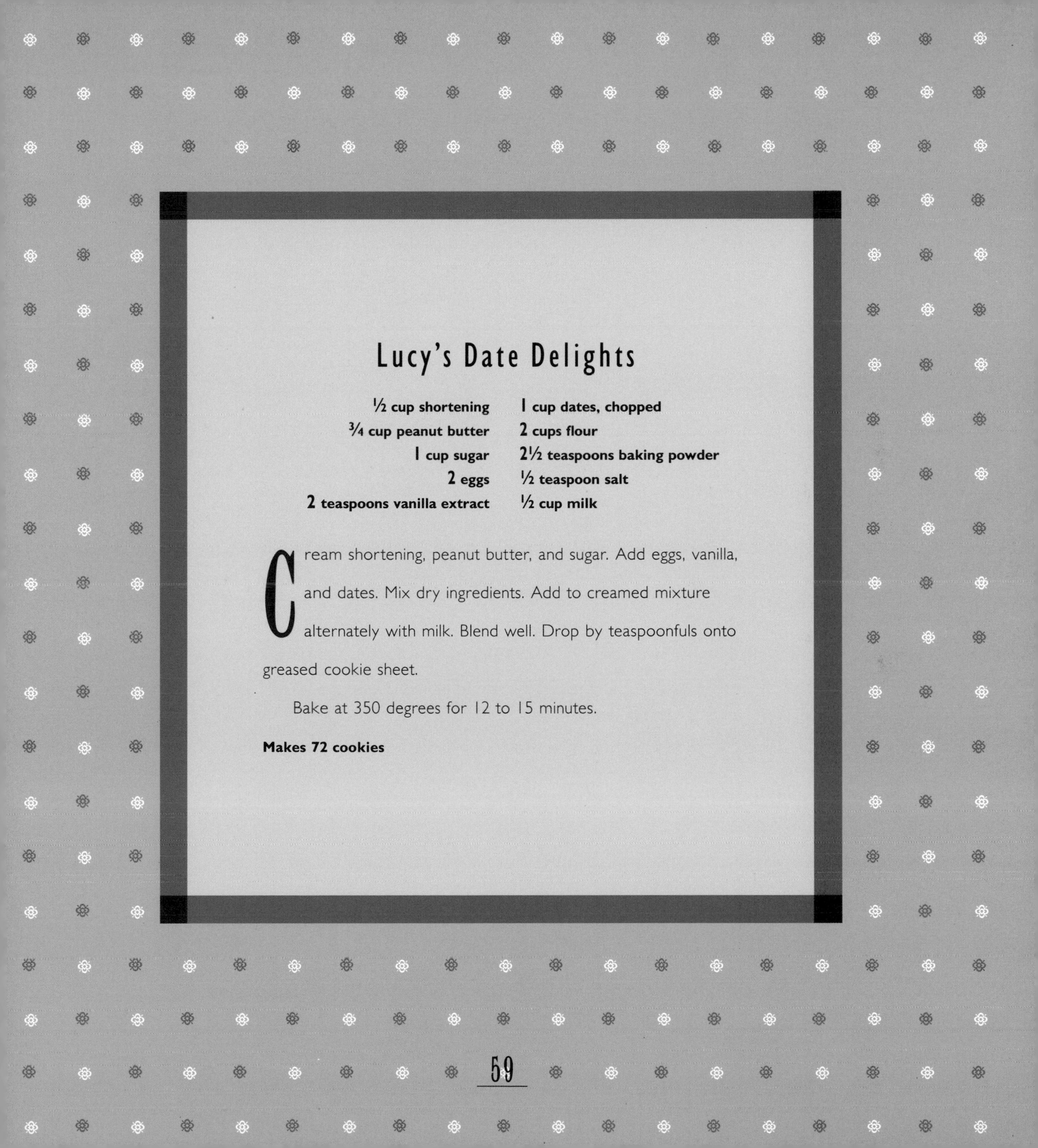

Lucy's Date Delights

½ cup shortening
¾ cup peanut butter
1 cup sugar
2 eggs
2 teaspoons vanilla extract
1 cup dates, chopped
2 cups flour
2½ teaspoons baking powder
½ teaspoon salt
½ cup milk

Cream shortening, peanut butter, and sugar. Add eggs, vanilla, and dates. Mix dry ingredients. Add to creamed mixture alternately with milk. Blend well. Drop by teaspoonfuls onto greased cookie sheet.

Bake at 350 degrees for 12 to 15 minutes.

Makes 72 cookies

Marcia's Apricot Delights

3 cups flour
1 tablespoon sugar
½ teaspoon salt
1 cup shortening
½ cup milk
1 package dry yeast
1 egg, slightly beaten
1 teaspoon vanilla extract
Confectioners' sugar
Apricot jam

Sift together flour, sugar, and salt. Cut in shortening until it resembles coarse crumbs. Heat milk to lukewarm and add package of yeast, slightly beaten egg, and vanilla. Add to flour mixture and form into a ball. Using a rolling pin sock, roll out ¼ of the dough at a time on a pastry cloth dusted with confectioners' sugar. With a pizza cutter, cut pastry into 2-inch squares. Put ½ tablespoon apricot jam in center of each square. Fold corners of pastry squares together and pinch. Let stand on cookie sheet 10 minutes before baking.

Bake at 350 degrees for 10 to 15 minutes.

Makes 60 cookies

Pottery Guild
Elsie the Cow
in Barrel

American Bisque
Head of a Majorette

Marla's Coconut Bars

CRUST

½ cup shortening	**½ teaspoon salt**
½ cup dark brown sugar, packed	**1 cup flour**

Cream shortening, brown sugar, and salt. Add flour and mix. Pat dough into bottom of 9 × 13 × 2″ pan.

Bake at 325 degrees for 20 minutes.

FILLING

2 eggs, well beaten	**½ teaspoon baking powder**
1 cup dark brown sugar, packed	**½ teaspoon salt**
1 teaspoon vanilla extract	**1½ cups coconut, shredded**
2 tablespoons flour	**1 cup nuts, chopped (optional)**

Combine all ingredients. Pour mixture over crust.

Bake at 325 degrees for 25 minutes. Cut bars while warm.

Makes 15 to 18 bars

Martha's Espresso Nut Meringues

4 egg whites
1 cup granulated sugar
½ teaspoon vinegar
1 teaspoon water
2 tablespoons instant espresso powder
2 teaspoons cornstarch
1 cup pecans, coarsely chopped

Whip egg whites until soft peaks form. Add sugar gradually, continuing to beat until meringue is stiff and glossy. Combine vinegar, water, espresso powder, and cornstarch. Stir until smooth and espresso is dissolved. Fold mixture into meringue. Fold in pecans. Drop by teaspoonfuls onto cookie sheet lined with foil.

Bake at 300 degrees for 30 to 35 minutes or until dry. Place in airtight container at once.

Makes 60 to 72 cookies

Martin's Meltaways

1 cup butter
½ cup confectioners' sugar
1 egg
2 ½ cups flour
½ cup walnuts, coarsely chopped
55 cloves, whole
Confectioners' sugar for rolling

Cream butter with confectioners' sugar. Add egg. Add flour and walnuts and mix well. Shape into round balls. Press one whole clove into each cookie.

On greased cookie sheet, bake at 350 degrees for 15 to 20 minutes or until golden. Roll in confectioners' sugar while warm.

Makes 55 cookies

Molly's Chocolate Sandwich Cookies

½ cup butter, softened
1 cup sugar
1 egg
1 teaspoon vanilla extract
1 cup milk
2 cups flour
1 teaspoon baking powder
1 teaspoon baking soda
½ teaspoon salt
½ cup cocoa

Cream butter and sugar. Add egg and vanilla. Add milk. Sift together dry ingredients and add to butter mixture. Drop by teaspoonfuls onto greased cookie sheet.

Bake at 350 degrees for 8 minutes.

FILLING

1 cup shortening
2 cups confectioners' sugar
1 cup Marshmallo Fluff®
¼ teaspoon milk
1 teaspoon vanilla extract

Mix all ingredients well. Cool and sandwich two cookies together with filling.

Makes 30 cookies

Hull
Little Red
Riding Hood

Fredricksburg Art
Cookie Shop
Windmill

Nuni's No-Bake Cookies

2 cups sugar
¼ cup butter or margarine
¼ cup cocoa
½ cup milk
3 cups uncooked quick rolled oats
½ cup peanut butter
1 teaspoon vanilla extract
Optional: 1 cup of any of the following: nuts, raisins, chocolate morsels, or shredded coconut

Stir sugar, butter, cocoa, and milk in a saucepan and cook until mixture comes to a good boil, stirring constantly. Boil for 1 minute. Pour over oats, peanut butter, and vanilla. Add optional ingredients. Drop by teaspoonfuls onto cookie sheet lined with foil and let cool.

Makes 40 cookies

Pat's Peanut Butter Fingers

½ cup butter
½ cup sugar
½ cup dark brown sugar
1 egg
⅓ cup peanut butter
½ teaspoon vanilla extract
½ teaspoon baking soda
¼ teaspoon salt
1 cup flour
1 cup quick cooking oats

Cream butter, sugar, and dark brown sugar well. Beat in egg, peanut butter, and vanilla. Sift together baking soda, salt, flour, and stir into mixture. Add oats. Spread batter into greased 9 × 13 × 2″ pan.

Bake at 350 degrees for approximately 20 minutes.

TOPPING

6-ounce package of semisweet chocolate morsels
½ cup confectioners' sugar, sifted
¼ cup peanut butter
4 tablespoons evaporated milk

Sprinkle warm cake with chocolate morsels and let stand 5 minutes. Spread softened chocolate evenly. Combine sugar, peanut butter, and evaporated milk and drizzle mixture over softened chocolate. Cool. Cut into fingers. Note: If doubling the recipe, use the same amount of chocolate morsels.

Makes 18 to 21 cookies

Six Cookie Jars
on Parade

COOKIE
TIME

National Silver
Chef

Ronnie's Orange Chocolate Drops

½ cup water
1½ cups sugar
2 tablespoons grated orange peel, candied
½ cup unsalted butter
1 egg
1 teaspoon vanilla extract
1 cup flour
½ cup unsweetened cocoa
2 teaspoons baking powder
½ teaspoon salt

Make a simple syrup of ½ cup water and ½ cup sugar. Add orange peel to syrup and boil for 8 minutes. Drain syrup from orange peel. Cream butter and 1 cup sugar. Add egg, vanilla, and orange peel. Combine remaining dry ingredients and add to creamed mixture. Mix well. Drop by teaspoonfuls onto ungreased cookie sheet.

Bake at 350 degrees for 12 minutes.

Makes 72 cookies

Robbinson Ransbottom
Smiling Oscar with Green Hat

Ruth's Almond Crescents

1 cup unsalted butter
3/4 cup confectioners' sugar
1 teaspoon vanilla extract
1 cup almonds, ground and unblanched
2 1/2 cups flour
Granulated sugar for rolling

Cream butter with confectioners' sugar and vanilla. Add ground almonds and flour and mix well. Shape dough into 1 1/2-inch crescents.

Bake at 375 degrees on ungreased cookie sheet for 20 minutes or until golden. Roll in granulated sugar.

Makes 60 cookies

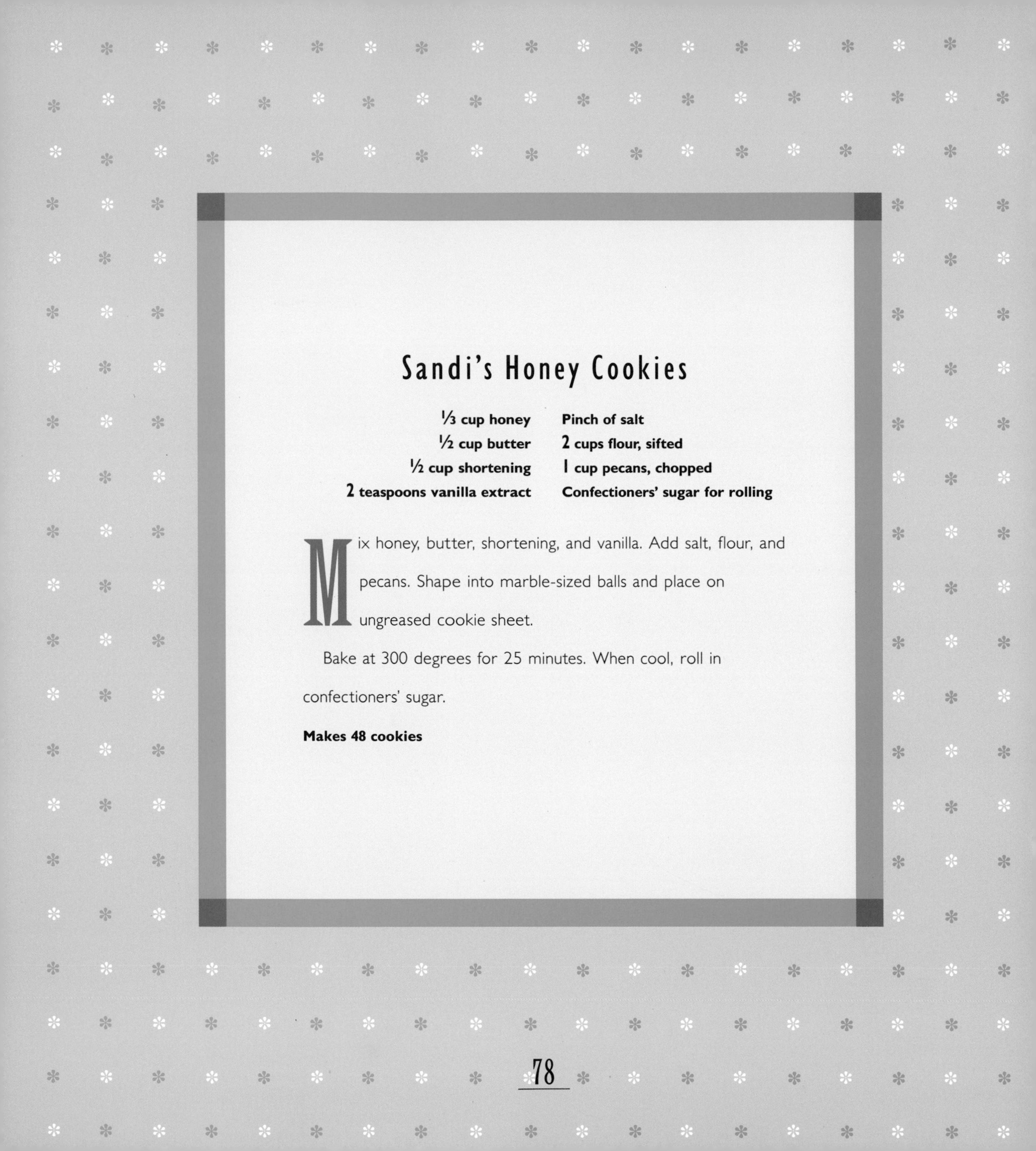

Sandi's Honey Cookies

⅓ cup honey
½ cup butter
½ cup shortening
2 teaspoons vanilla extract
Pinch of salt
2 cups flour, sifted
1 cup pecans, chopped
Confectioners' sugar for rolling

Mix honey, butter, shortening, and vanilla. Add salt, flour, and pecans. Shape into marble-sized balls and place on ungreased cookie sheet.

Bake at 300 degrees for 25 minutes. When cool, roll in confectioners' sugar.

Makes 48 cookies

Robbinson Ransbottom
Cow Jumped Over the Moon

Smiling Pig Wearing Beanie

Seth's Brandy Delights

1 cup unsalted butter
½ cup confectioners' sugar
1 egg
1 teaspoon vanilla extract
2 tablespoons brandy
2¼ cups flour
¾ cup walnuts, ground

Cream butter, sugar, egg, vanilla, and brandy. Blend in flour and nuts. Shape dough into marble-sized balls and place on ungreased cookie sheet.

Bake at 375 degrees for 10 to 12 minutes.

Makes 48 cookies

Sloane's Four-Way Sables

2 cups butter
1¼ cups sugar
4½ cups flour
½ cup chopped raisins soaked in ¼ cup rum
½ teaspoon lemon extract with grated rind of 1 lemon
½ cup toasted coconut with ½ teaspoon vanilla extract
½ cup almonds, blanched, toasted, and coarsely chopped, with ½ teaspoon almond extract
Granulated sugar for rolling

Cream butter and sugar. Add flour and mix well. Divide dough into four parts. Add raisins to one part, lemon extract to one part, toasted coconut to one part, and toasted almonds to one part. Roll each portion of dough into a 1½-inch log. Wrap in wax paper and chill logs until firm. Roll logs in granulated sugar. Slice and place on greased cookie sheet.

Bake at 350 degrees for 10 to 15 minutes.

Makes 40 cookies

Brush
Happy Squirrel with Top Hat

Stacey's Chocolate Chunks

3 cups semisweet or bittersweet chocolate, chopped
4 ounces unsweetened chocolate
½ cup unsalted butter
½ cup flour
½ teaspoon baking powder
⅛ teaspoon salt
4 extra large eggs
1½ cups sugar
1 tablespoon vanilla extract
2 tablespoons ground espresso beans or instant espresso powder

Melt 1½ cups of the semisweet chocolate, the unsweetened chocolate, and butter in double boiler. Cool to room temperature. Mix flour, baking powder, and salt. Beat eggs and sugar until thick. Beat in vanilla and espresso. Gently fold in melted chocolate and then dry ingredients. Add remaining 1½ cups of chopped semisweet chocolate. Refrigerate 1 hour. Drop by teaspoonfuls onto greased cookie sheet.

Bake at 350 degrees for 8 minutes. Remove when edges are crisp and cookies start to crack.

Makes 40 cookies

Stephen's Glazed Lemon Rounds

¾ cup unsalted butter, at room temperature
¼ teaspoon almond extract
½ teaspoon lemon extract
1 cup flour
½ cup confectioners' sugar
1 tablespoon cornstarch
¼ teaspoon salt

Cream butter until light. Add almond and lemon extracts. Sift together dry ingredients and add to butter mixture. Blend thoroughly. Divide dough into two equal portions and roll into logs on surface dusted with confectioners' sugar. Refrigerate until firm. Slice dough about ⅛-inch thick. Place about 1 inch apart on lightly greased cookie sheet.

Bake at 350 degrees for 10 minutes or until golden brown.

GLAZE

⅔ cup confectioners' sugar
4 tablespoons fresh lemon juice, strained

Mix the two ingredients together. Glaze when cookies are cool.

Makes 36 cookies

Dog with Cookies Cookbook

Thor's Thumbprint Cookies

¾ cup butter, at room temperature
¾ cup confectioners' sugar
1 egg yolk
½ teaspoon almond extract
1¾ cups flour
½ cup jam or jelly

Cream butter and sugar well. Add yolk of egg and mix thoroughly. Add almond extract and flour. Mix well. Refrigerate wrapped dough for 2 hours. Form marble-sized balls of dough. Place on greased cookie sheet. Press thumb into middle of each ball. Fill with jam or jelly.

Bake at 350 degrees for 10 to 12 minutes.

Makes 36 cookies

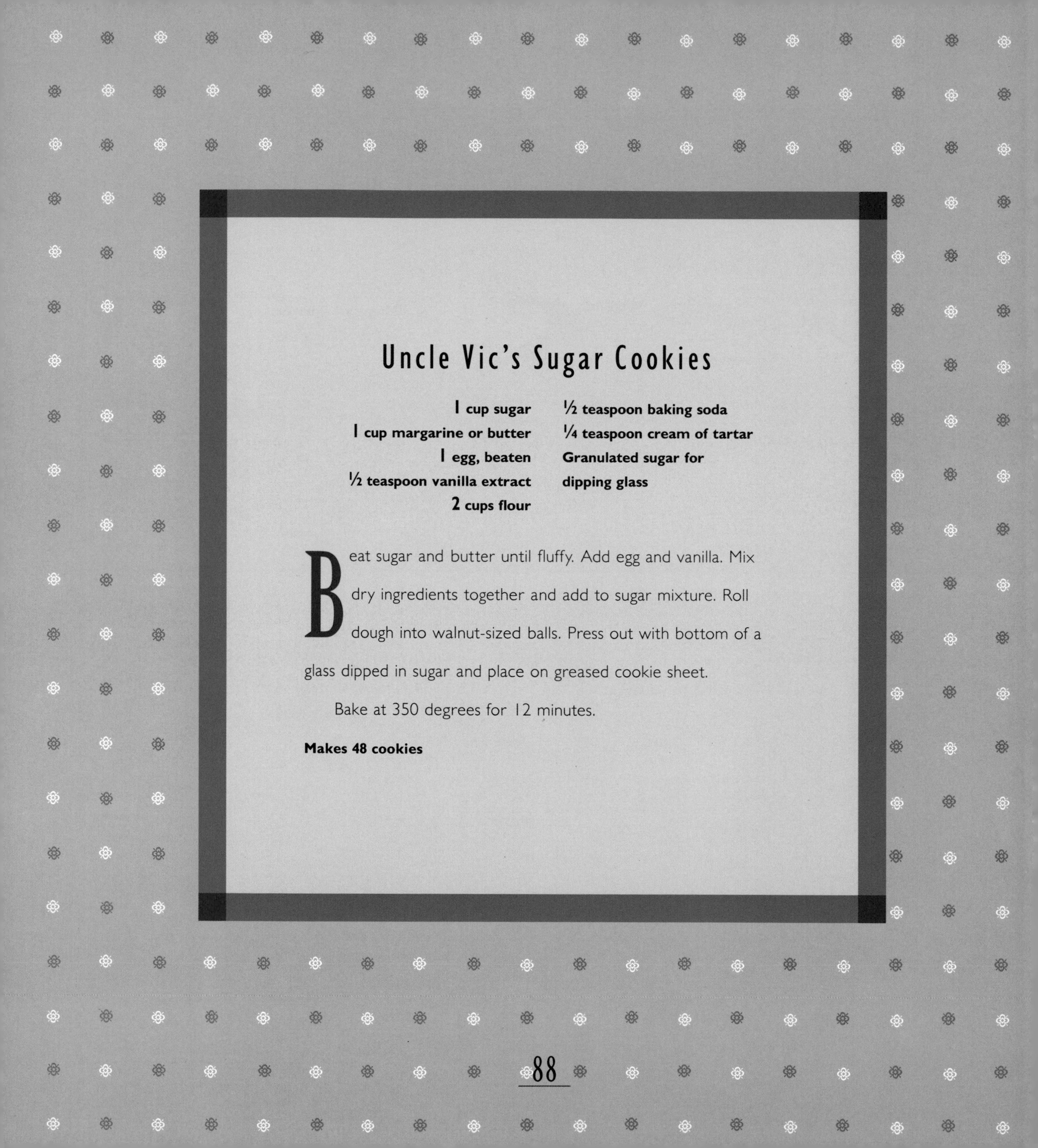

Uncle Vic's Sugar Cookies

1 cup sugar
1 cup margarine or butter
1 egg, beaten
½ teaspoon vanilla extract
2 cups flour
½ teaspoon baking soda
¼ teaspoon cream of tartar
Granulated sugar for dipping glass

Beat sugar and butter until fluffy. Add egg and vanilla. Mix dry ingredients together and add to sugar mixture. Roll dough into walnut-sized balls. Press out with bottom of a glass dipped in sugar and place on greased cookie sheet.

Bake at 350 degrees for 12 minutes.

Makes 48 cookies

Shawnee
"Muggsy" Terrier
with Blue Bow,
White Pup with
Blue Bow

Shawnee

Winking Owls

Victoria's Ginger Cookies

1 egg, beaten
1 cup sugar
¾ cup half shortening, half butter
4 tablespoons molasses
1 teaspoon cinnamon
1 teaspoon ground ginger
½ teaspoon salt
2½ teaspoons baking soda
2 cups flour
Granulated sugar for rolling

Cream egg, sugar, shortening mixture, and molasses. Sift together dry ingredients and fold into creamed mixture. Chill 1 hour. Shape into marble-sized balls. Roll in sugar and place on ungreased cookie sheet.

Bake at 375 degrees for 10 minutes.

Makes 48 cookies

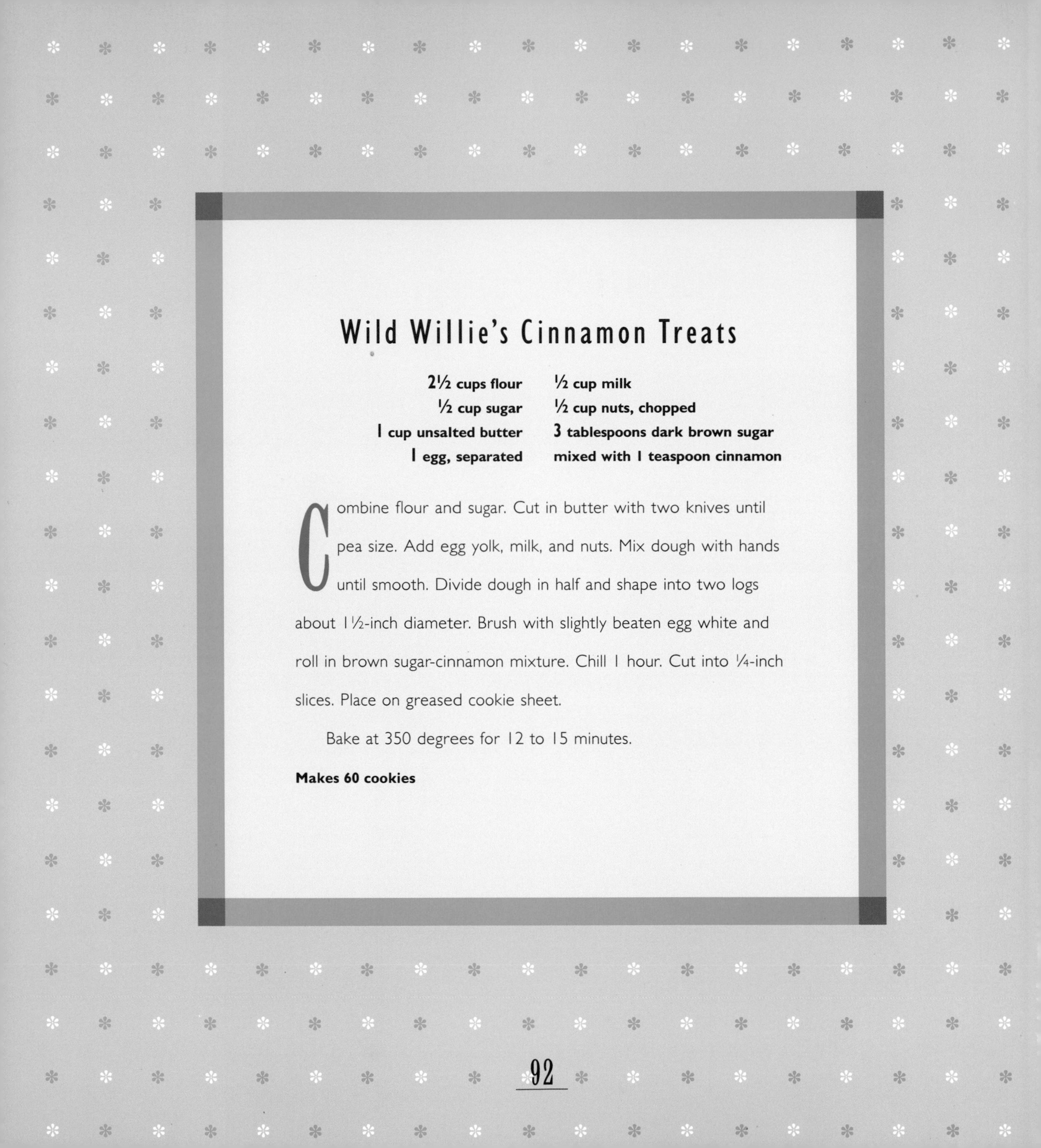

Wild Willie's Cinnamon Treats

2½ cups flour
½ cup sugar
1 cup unsalted butter
1 egg, separated
½ cup milk
½ cup nuts, chopped
3 tablespoons dark brown sugar mixed with 1 teaspoon cinnamon

Combine flour and sugar. Cut in butter with two knives until pea size. Add egg yolk, milk, and nuts. Mix dough with hands until smooth. Divide dough in half and shape into two logs about 1½-inch diameter. Brush with slightly beaten egg white and roll in brown sugar-cinnamon mixture. Chill 1 hour. Cut into ¼-inch slices. Place on greased cookie sheet.

Bake at 350 degrees for 12 to 15 minutes.

Makes 60 cookies

American Bisque
Boy and Girl Under an Umbrella

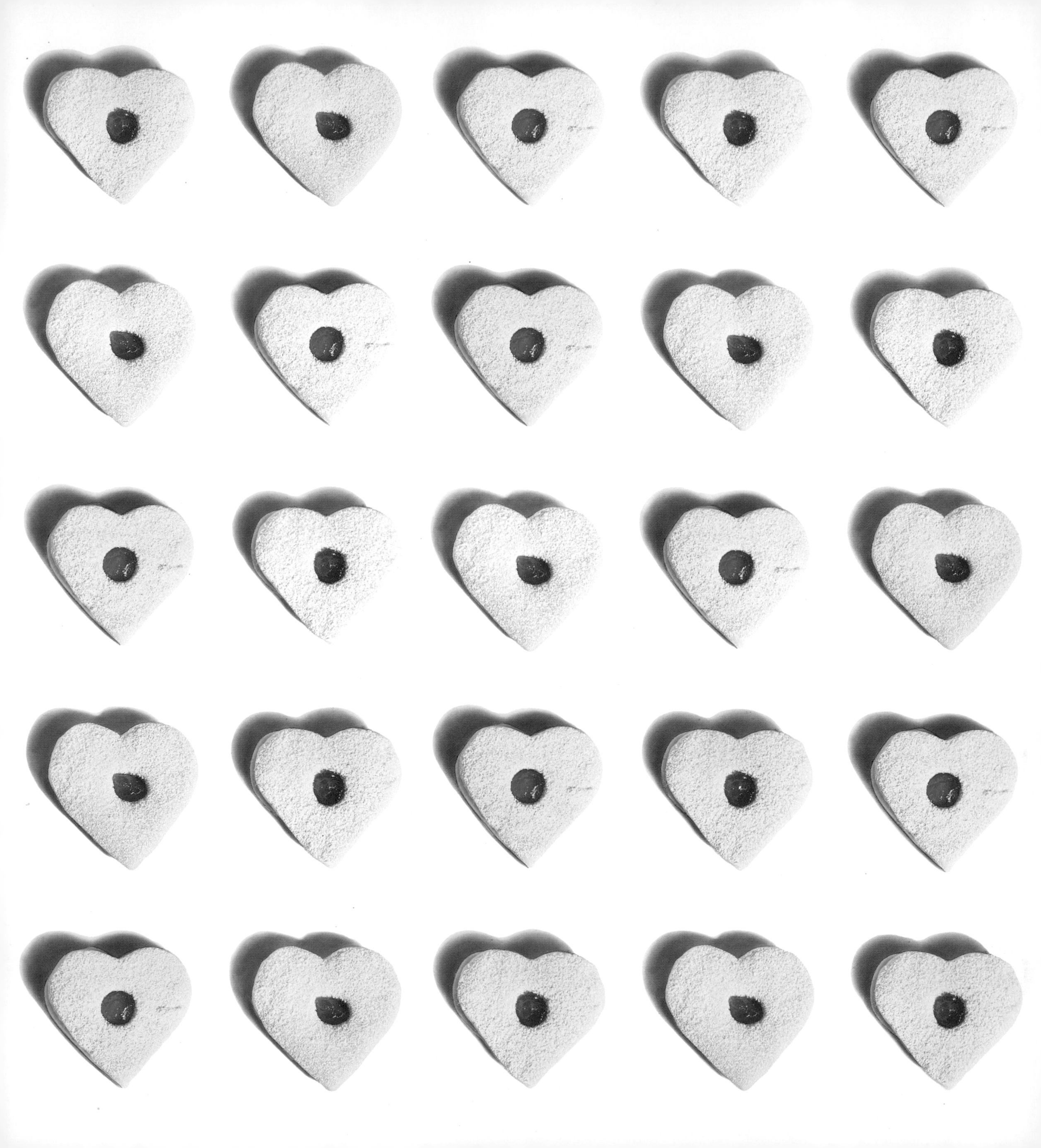

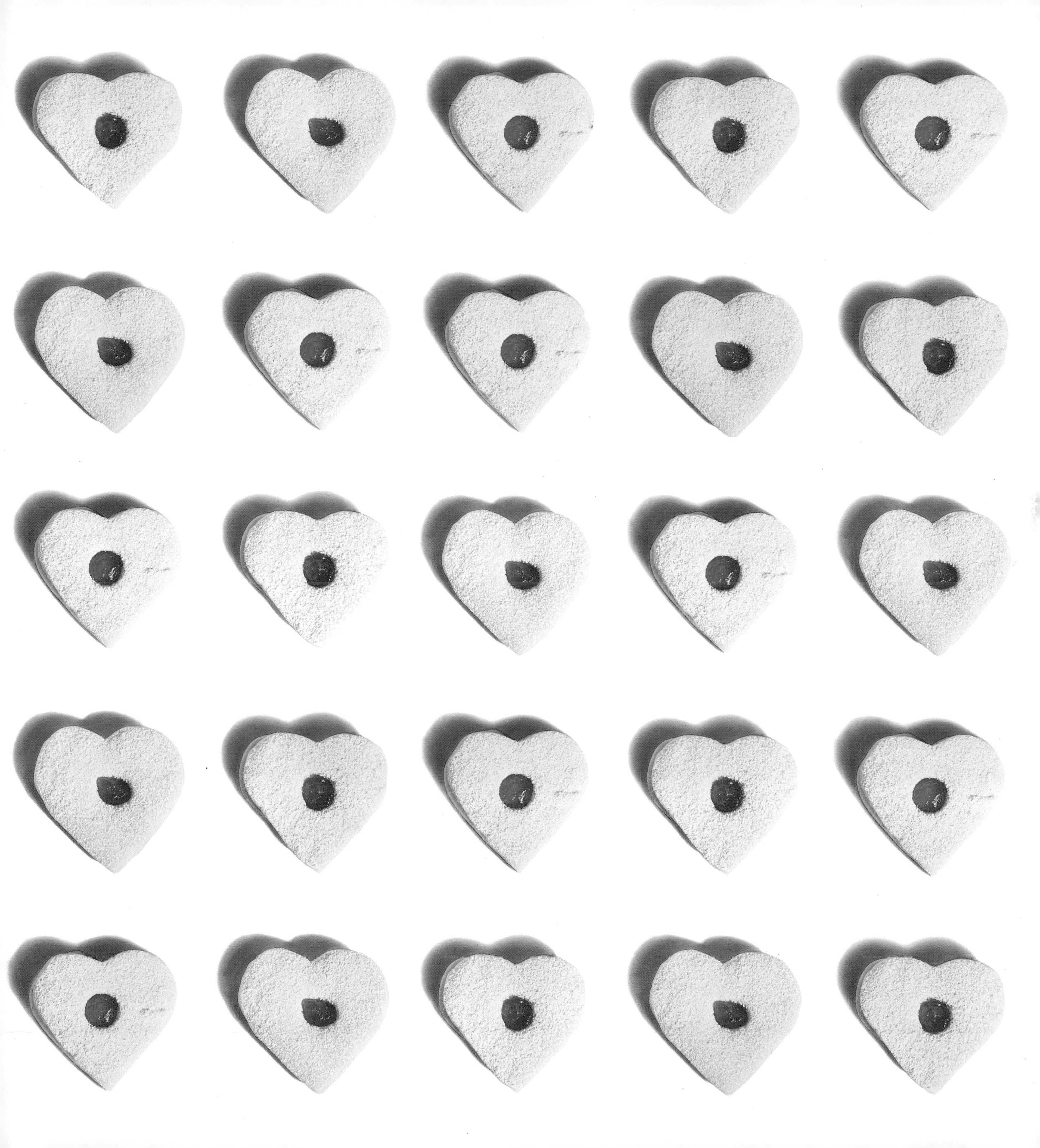

Rabbit in a Hat